PASSWORD KEEPER:

An Old-Fashioned Logbook to Protect My Identity From Those Who Would Steal It!

Date Started:

Date: Finished:

scribble your name:

Address:

Tiffany Inks Studio LLC

An artist and writer's

c o p y r i g h t

encourages a culture of support

and honors the person

who creates, writes, & designs

publications & works of art.

Cover design, artwork, and interior formatting by Patricia Tiffany Morris, owner of Tiffany Inks Studio LLC.

Tiffany Inks Studio LLC
PO Box 299
Bondurant, IA 50035

ISBN: 978-1-95527-4-02-9

Endorsements:

"An inspiration station for contemplators, doodlers, journalers, planners, dreamers, and creatives of many kinds who want to renew thoughts and cultivate gratitude in a hands-on way. Patricia inspires and guides with a variety of ways to combine art, bible study, prayer, contemplation, dreaming, and goal setting all based on Philippians 4:8.

I particularly love that she's provided a little help for those who are just getting started with adding lettering and drawing in their spiritual practice."

~**Ginger Harrington :** Author of **Holy in the Moment** (GingerHarrington.com)

"With breathtaking beauty, Patricia Tiffany Morris has created page after page of prompts that invite you to open your mind and pour out, gather, and organize thoughts, plans, feelings, and details of your life. The stunning pages in Journaling Scribbles elevate my mind-musings to a whole new level that compliments the busy neurology going on, even in my dreams.

Studies show the act of writing, gripping a utensil, and tapping into our inner world, spilling out our thoughts, activates parts of our brain that reduce stress, increases wellbeing, organizes the chaos of uncertainty, and increases our ability to reach our goals.

What an incredibly beautiful way Patricia Tiffany Morris has created to help us "mind-map" what we think... and show us who we are. My thoughts never looked so good! What a wonderful thing to pick these books up in years to come and say, "Yes, that was me!"

~**Deborah McCormick Maxey, PhD:** Licensed Professional Counselor, Licensed Marriage and Family Therapist , and Author of **The Endling** (DeborahMaxey.com/books)

"If you are looking for something to jumpstart your journaling process, Patricia Tiffany Morris has the answer in this series of journals. Beautifully illustrated with Patricia's own art, you're sure to discover a creative way to organize and process your busy days in a way that fits your lifestyle."

~**Mary Potter Kenyon :** Program Coordinator for Shalom Spirituality Center, certified grief counselor, Therapeutic Art Coach, and author of seven books, including **Called to Be Creative:** A Guide to Reigniting Your Creativity (MaryPotterKenyon.com)

"As you open this book you realize; each page is an invitation to dance. Patricia Tiffany Morris is a wonder. Since the moment we met, I've been constantly astounded by her capacity to pour out her zeal to encourage and build others up. Some might be content to create whimsical and inspiring art, and stunningly beautiful poetry and prose, then sit back and enjoy the accolades and awards that go along with it.

Not Patricia. Sure, she's got those too. But she's busily passing out keys to the Kingdom in every way she can, that our own art, writing, and our relationship with Jesus, might flourish. Social media trainings. Tech trainings. Biblical devotionals. Now, her creativity and planning journals, diaries, & sketchbooks.

I strongly recommend you snap up an armful for all the dreamers and the doers in your life. And for all you artists, here's the really unique thing about these journals: Patricia knows how the creative mind works and celebrates it! Well done, Patricia."

~**Wendy Hibbard :** Founder, The Writing Room and SHP Coaching; Editor, **Beneath the Mask:** Faith, Hope & Transformation in the Face of COVID-19 (wendyhibbard.com)

to the One who created all things and for His glory.

artist & creator of journaling scribbles:

Find me on social-media and share a page from your journal

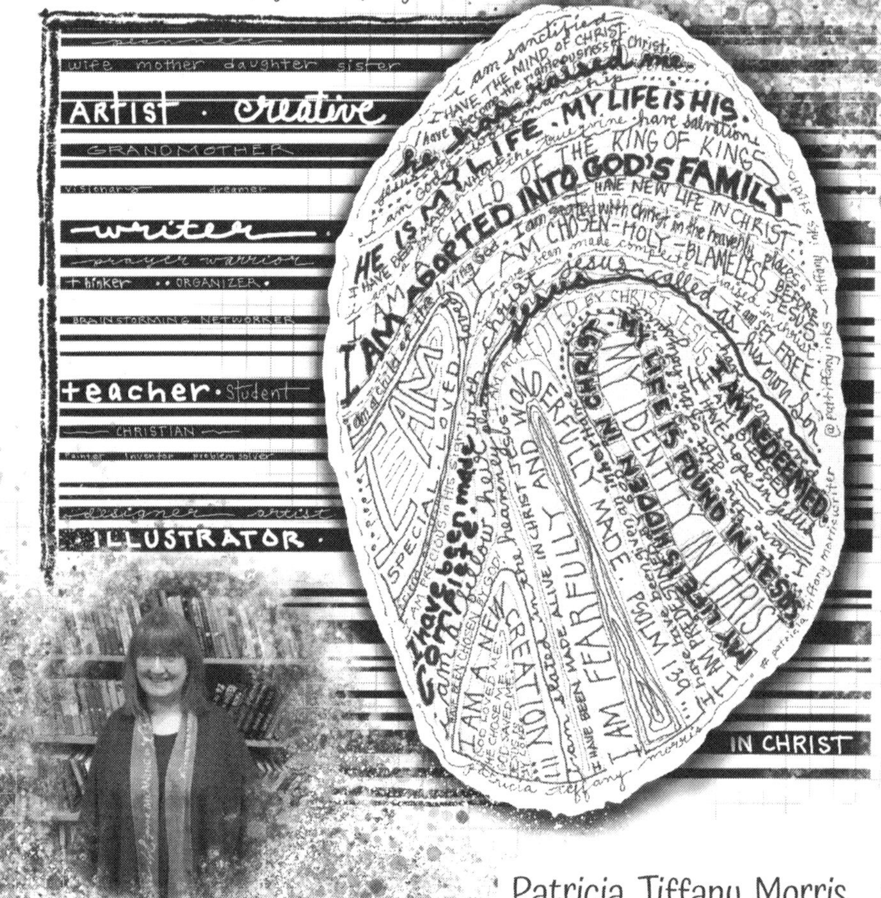

Patricia Tiffany Morris

www.Patriciatiffanymorris.com

www.pinterest.com/patriciatiffanymorris.com
www.youtube.com/channel/UC8CHWtUkQcCJwbjBOAnTRKg
www.facebook.com/TiffanyInksStudio/
www.instagram.com/PatriciaTiffanyMorris/

Thank you for your purchase!

I'm grateful you chose this resource and pray you are blessed and encouraged. If you'd like to send me a photo of a page from your journal, I'd love to see what you create! I'll be sharing some of your designs on my blog. Please share your pages with me at: **www.patriciatiffanymorris.com**

where to find the artist:

Artist, Writer, and Christian Creative, Patricia Tiffany Morris, weaves original hand drawn artwork, digital illustrations, and traditional planning elements throughout the pages.

You can find her on Instagram FB, IG, YouTube, or Pinterest sharing her coloring pages, procreate time lapse videos, sunrise and sunset pictures, artwork, and writing poetry, prose, or short stories. An aspiring children's writer and illustrator, Patricia looks forward to new projects during retirement years.

more about the artist:

Patricia Tiffany Morris sketches ideas in her sleep, that is when she finds time to sleep. As a young child she read late into the night, her mom reminding her she would have school in a few hours. Staying up all night served her well during architecture studio projects at ISU in the 80's.

After marriage to her park ranger-husband. she home educated her five daughters, immersing them in the imaginative worlds of Little Women, Little House on the Prairie, and Nancy Drew. Patricia's favorite authors include classics by John Bunyan, C.S. Lewis, A.A. Milne, and Dr. Seuss.

She actively supports fellow authors. Teaching art and computer organization and skills, Patricia offers other geeky tech services through Tiffany Inks Studio LLC.

Now an empty-nester, Patricia pursues both writing and art. Inspired by her humorous, rhyming husband, she brainstorms new stories as he reads her writing & poetry in delightful character voices.

She works full-time as the owner of Tiffany Inks Studio LLC, publisher of the JOURNALING SCRIBBLES SERIES.

Journaling Scribbles™

Tiffany Inks Studio LLC

Color, scribble, sketch, draw, & write ideas on the following pages.

visual table of contents

Use drawings, sketches, or words to organize your pages.

visual table of contents

Sketch your ideas for your planner or list pages & contents.

create a key

test your tools & colors

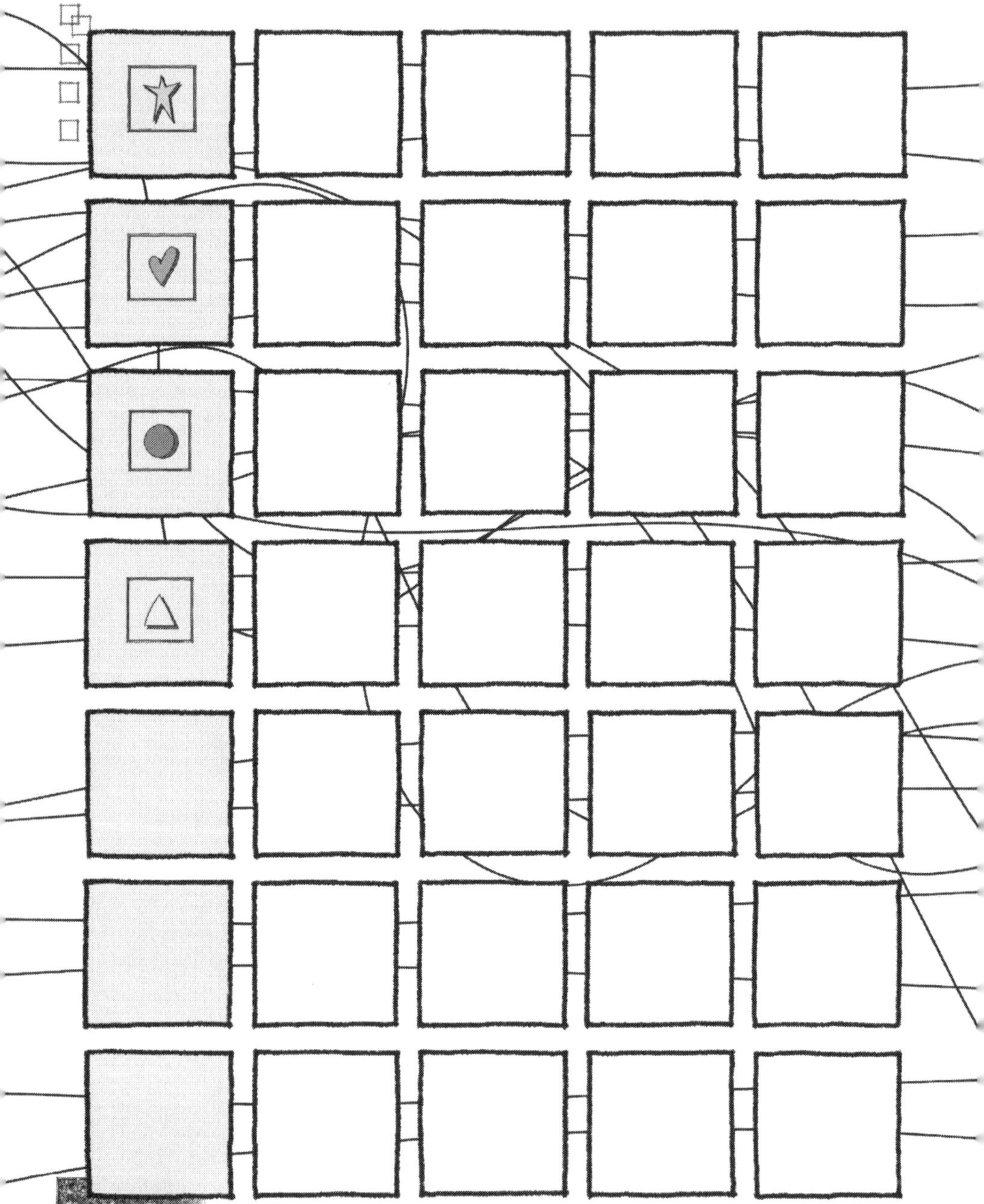

create a palette

test your tools & colors

vision board

Write words to inspire your journey through the pages of this journal, and creatively letter or color your ideas.

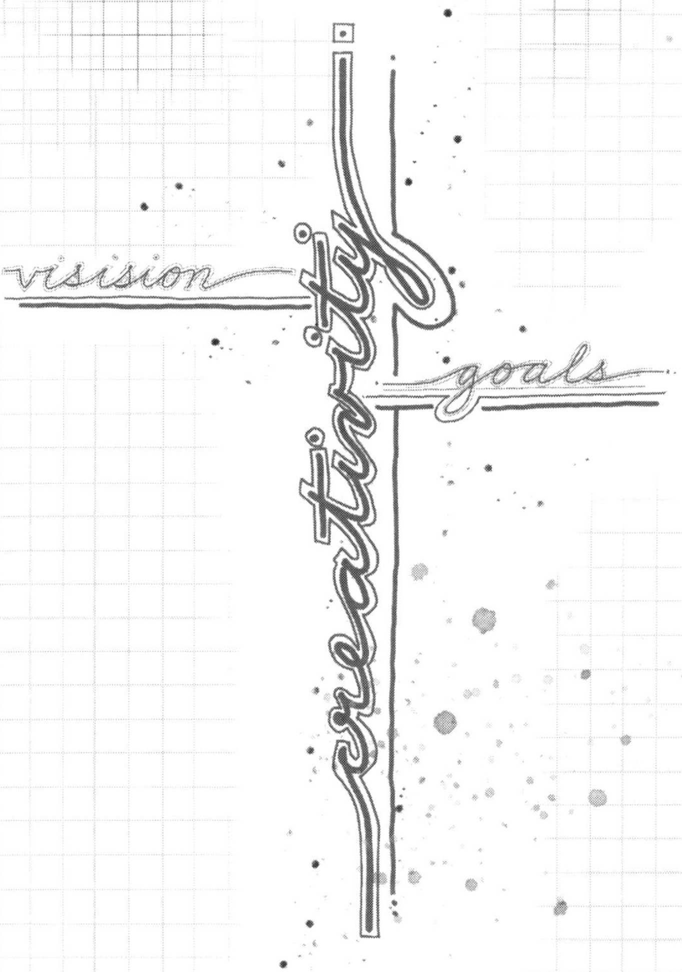

goals & objectives

Write your goals & objectives here. Scribble ideas in the margins.

goal #1

1
2
3
4
5
6
7

goal #2

1
2
3
4
5
6
7

goal #3

1
2
3
4
5
6
7

INTENTIONALITY WITH CREATIVITY

The key to organizing & managing anything is often elusive. But add a little color & creativity, and find peace in spite of the chaos of daily life.

Ideas for organizing your home and work spaces:

1. Use a color-coded key to indicate status and priority or type of project.
2. Which colors provide visual emphasis and "at-a-glance" focus for you?
3. Take time to gather your pens, tools, and colored pencils. List them.
4. Choose 5-7 favorite colors or use the colors matching the cover art.
5. Organizing uses the left side of the brain. Plan, Order, Numbering, Listing, etc.
6. Coloring uses the right side of the brain. Creativity, Art, Doodling, Coloring, etc.
7. Which type of left-brain tasks help you organize?
8. Which type of right-brain activities help you enjoy organizing?
9. How else can you help create a peaceful and productive home & work space?

PLANNING FOR SECURITY

Su M T W T F Sa

lists & more lists

ADD NEW ACCOUNTS & UPDATE EXISTING

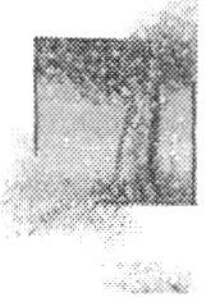

Updating your online account user names and passwords, and security questions takes time. Set manageable goals and track just 15 minutes a day to add content to this logbook.

List your accounts below. Check off which days you access them most. Put the MOST USED accounts at the front sections of this book.

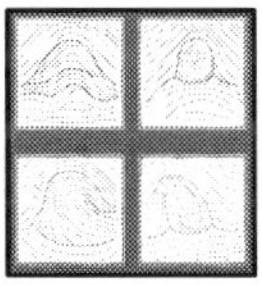

☐	Account	Su	M	T	W	T	F	Sa
☐		☐	☐	☐	☐	☐	☐	☐
☐		☐	☐	☐	☐	☐	☐	☐
☐		☐	☐	☐	☐	☐	☐	☐
☐		☐	☐	☐	☐	☐	☐	☐
☐		☐	☐	☐	☐	☐	☐	☐
☐		☐	☐	☐	☐	☐	☐	☐
☐		☐	☐	☐	☐	☐	☐	☐
☐		☐	☐	☐	☐	☐	☐	☐
☐		☐	☐	☐	☐	☐	☐	☐
☐		☐	☐	☐	☐	☐	☐	☐
☐		☐	☐	☐	☐	☐	☐	☐
☐		☐	☐	☐	☐	☐	☐	☐
☐		☐	☐	☐	☐	☐	☐	☐
☐		☐	☐	☐	☐	☐	☐	☐
☐		☐	☐	☐	☐	☐	☐	☐
☐		☐	☐	☐	☐	☐	☐	☐
☐		☐	☐	☐	☐	☐	☐	☐
☐		☐	☐	☐	☐	☐	☐	☐

THE PASSWORD KEEPER

Keep this book secure at all times!

AN OLD-FASHIONED LOGBOOK TO PROTECT MY IDENTITY FROM THOSE WHO WOULD STEAL IT!

THIS BOOK BELONGS TO:

PROS & CONS OF A HANDWRITTEN LOG

List a few challenges or objections to writing your passwords in an old-fashioned logbook instead of storing them online in a digital app.

List some of the reasons to take your passwords offline.

- [ ] ______
- [ ] ______
- [ ] ______
- [ ] ______
- [ ] ______
- [ ] ______
- [ ] ______
- [ ] ______
- [ ] ______

Su M T W T F Sa

PROTECT YOUR IDENTITY

MY IDENTITY

Write whatever is unique about you.
Write in the space between the lines.

I AM ...

When you're finished, try the word art on the next page.

FINGERPRINT DOODLING ART

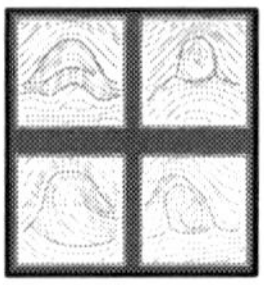

Write whatever is unique about you.
Write in the space between the lines.

WHO DOES GOD SAY YOU ARE?

#ID

FINGERPRINT IDENTIFICATION

MY IDENTITY

List important facts about your ID.

1. Make passwords unique!

2. Don't answer security questions with answers that are true. Be Creative.

3. ______

4. ______

5. ______

6. ______

7. ______

8. ______

9. ______

10. ______

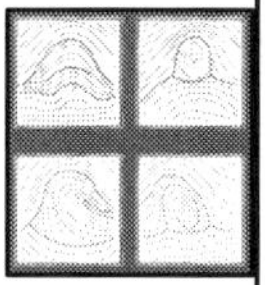

Su	M	T	W	T	F	Sa

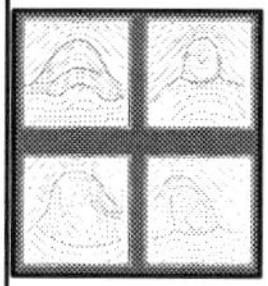

BRAINSTORMING

What categories can you think of to organize your passwords and make them easier to locate?

Color, scribble, sketch, draw, & write throughout the pages of this journal.

LIST YOUR ELECTRONIC DEVICES

List the various devices you own. Devote one page to each device.

On the following pages you will organize your log in information.

Don't forget to list printers, routers, modems, scanners, etc.

1

2

3

4

5

6

7

8

9

10

11

12

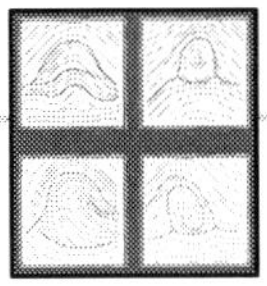

SECURITY CONCERNS

What else concerns you?

Passwords need to be 8-24 characters. Use a variety of lower and upper case letters and 1-2 special characters. Many keys on your keyboard can be used. But some places restrict the actual keys you can use.

Hackers and their nefarious apps work all year long world-wide. Don't make their job easy. Try some unusual combinations of keys. Don't use same letter more than 2 times in a row.

Criminals can use your social media to target possible password trends (sports, religion, cars, etc) Don't use common names or dates in your passwords.

Minimize the use of same password on multiple applications. Mix it up and keep your passwords distinct on every account.

Don't share your passwords. Keep this book secure and make a copy for your local box. Perhaps add this book to your legal papers for ease of your family to manage your estate.

LEGACY & ESTATE

Any other ideas?

Do you have a file prepared with the locations of your password book or access to your online accounts?

Who will be able to find your passwords and manage or close your accounts if you become incapacitated?

Make a copy of this book and put it into a safe deposit box or in your estate planner. Deliver a copy to your attorney or trust.

Highlight or mark password sections that your spouse, family, or executor might need when you are gone or in the hospital.

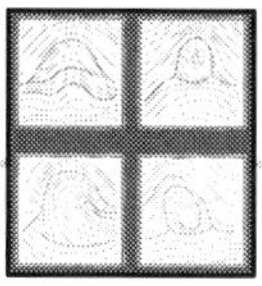

10 PASSWORD KEEPER TIPS

NAME DATE NOTES

USER

EMAIL

PW

IDEAS FOR KEEPING PASSWORDS SECURE

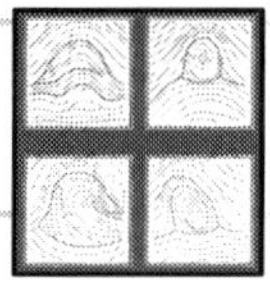

1. Organize passwords by type or alphabetical.
2. Add labels at each section.
3. Keep this book with you & a copy in a secure place with instructions for your family.
4. Write the date when you opened each account. This is important when trying to recover accounts.
5. Add security questions in notes section.
6. Use pencil for passwords so you can change the password regularly. Most other items don't change very often.
7. Shorten the actual email in case someone finds this book that you don't want to have the complete email.
8. Don't write your full email ANYWHERE in this book. Abbreviate it if 2 or more emails.
9. Write a phone or contact name in the notes.
10. Some accounts have a user name different from your email.

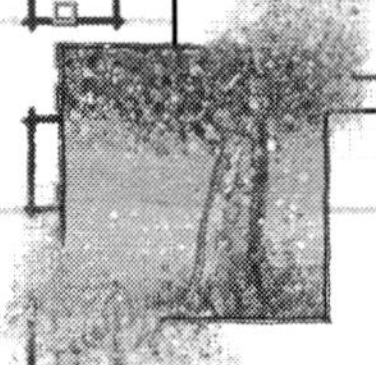

INSTRUCTIONS FOR EACH ENTRY

NAME & DATE: WRITE THE NAME OF THE ACCOUNT & THE DATE YOU OPENED IT.

USERNAME: SOME ACCOUNTS ASK FOR A SEPARATE USERNAME.

EMAIL: ABBREVIATE YOUR EMAIL. DON'T WRITE ACTUAL ACCOUNT NAME.

PASSWORD: WRITE YOUR UNIQUE PASSWORD. USE PENCIL TO CHANGE IT LATER.

SOME ACCOUNTS HAVE SECURITY PINS.

SOME ACCOUNTS HAVE SECURITY QUESTIONS.

SOME ACCOUNTS HAVE 2-STEP VERIFICATION.

SOME ASK YOU TO ADD A SECONDARY PHONE NUMBER.

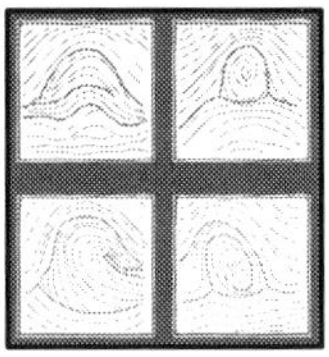

PASSWORD CATEGORIES

Internet, Tech Guy, Router, and Modem

Computer, Laptop, Printers, and Scanners

Phones, Emails for Personal and Business

Banking, Credit Cards, Financials, and Estate

App Store, Windows Store, iTunes, and Specific Apps

Cloud Storage, One Drive, and Software by Type

Website, Blog, Etsy, Square, and Online Stores

IRS, State Treasurer, Government, and State Offices

USPS, Shipping, Office Supplies, & Memberships

Cable, Subscriptions, and Entertainment

Social Media, FB, IG, Twitter, Gab, Pinterest, and More

Zoom, Skillshare, Udemy, and Teaching Platforms

Shopping, Sellers, Groceries, and Online Purchasing

Utilities, Mortgage, Real Estate, and House Repairs

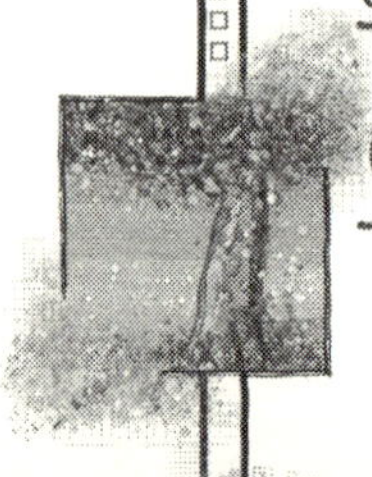

Create New Categories for your personal needs

LIST YOUR CATEGORIES

PASSWORD GROUP

NAME | DATE | NOTES

USER

EMAIL

PW

NAME | DATE | NOTES

USER

EMAIL

PW

NAME | DATE | NOTES

USER

EMAIL

PW

NAME | DATE | NOTES

USER

EMAIL

PW

PASSWORD GROUP

NAME DATE NOTES

USER

EMAIL

PW

NAME DATE NOTES

USER

EMAIL

PW

NAME DATE NOTES

USER

EMAIL

PW

NAME DATE NOTES

USER

EMAIL

PW

PASSWORD GROUP

NAME DATE NOTES

USER

EMAIL

PW

NAME DATE NOTES

USER

EMAIL

PW

NAME DATE NOTES

USER

EMAIL

PW

NAME DATE NOTES

USER

EMAIL

PW

PASSWORD GROUP

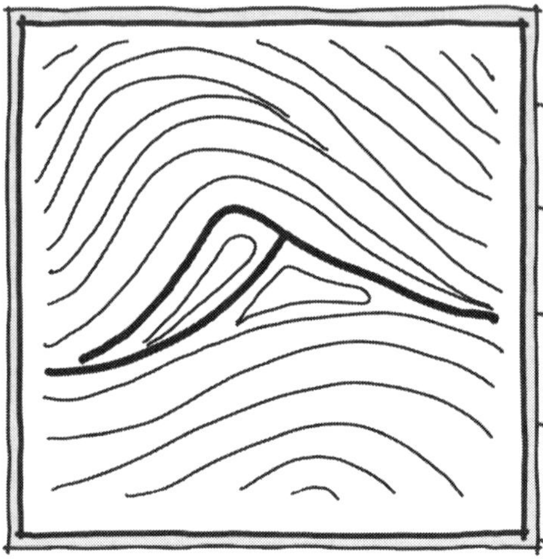

NAME DATE NOTES

USER

EMAIL

PW

NAME DATE NOTES

USER

EMAIL

PW

NAME DATE NOTES

USER

EMAIL

PW

NAME DATE NOTES

USER

EMAIL

PW

NAME DATE NOTES

USER

EMAIL

PW

NAME DATE NOTES

USER

EMAIL

PW

NAME DATE NOTES

USER

EMAIL

PW

NAME DATE NOTES

USER

EMAIL

PW

NAME DATE NOTES

USER

EMAIL

PW

NAME DATE NOTES

USER

EMAIL

PW

NAME DATE NOTES

USER

EMAIL

PW

NAME DATE NOTES

USER

EMAIL

PW

PASSWORD GROUP

NAME DATE NOTES

USER

EMAIL

PW

NAME DATE NOTES

USER

EMAIL

PW

NAME DATE NOTES

USER

EMAIL

PW

NAME DATE NOTES

USER

EMAIL

PW

NAME DATE NOTES

USER

EMAIL

PW

NAME DATE NOTES

USER

EMAIL

PW

NAME DATE NOTES

USER

EMAIL

PW

NAME DATE NOTES

USER

EMAIL

PW

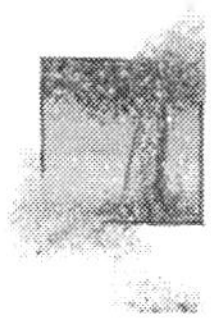

NAME DATE NOTES

USER

EMAIL

PW

NAME DATE NOTES

USER

EMAIL

PW

NAME DATE NOTES

USER

EMAIL

PW

NAME DATE NOTES

USER

EMAIL

PW

PASSWORD GROUP

NAME DATE NOTES

USER

EMAIL

PW

NAME DATE NOTES

USER

EMAIL

PW

NAME DATE NOTES

USER

EMAIL

PW

NAME DATE NOTES

USER

EMAIL

PW

NAME DATE NOTES

USER

EMAIL

PW

NAME DATE NOTES

USER

EMAIL

PW

NAME DATE NOTES

USER

EMAIL

PW

NAME DATE NOTES

USER

EMAIL

PW

NAME DATE NOTES

USER

EMAIL

PW

NAME DATE NOTES

USER

EMAIL

PW

NAME DATE NOTES

USER

EMAIL

PW

NAME DATE NOTES

USER

EMAIL

PW

PASSWORD GROUP

NAME DATE NOTES

USER

EMAIL

PW

NAME DATE NOTES

USER

EMAIL

PW

NAME DATE NOTES

USER

EMAIL

PW

NAME DATE NOTES

USER

EMAIL

PW

NAME DATE NOTES

USER

EMAIL

PW

NAME DATE NOTES

USER

EMAIL

PW

NAME DATE NOTES

USER

EMAIL

PW

NAME DATE NOTES

USER

EMAIL

PW

NAME DATE NOTES

USER

EMAIL

PW

NAME DATE NOTES

USER

EMAIL

PW

NAME DATE NOTES

USER

EMAIL

PW

NAME DATE NOTES

USER

EMAIL

PW

NAME DATE NOTES

USER

EMAIL

PW

NAME DATE NOTES

USER

EMAIL

PW

NAME DATE NOTES

USER

EMAIL

PW

NAME DATE NOTES

USER

EMAIL

PW

NAME DATE NOTES

USER

EMAIL

PW

NAME DATE NOTES

USER

EMAIL

PW

NAME DATE NOTES

USER

EMAIL

PW

NAME DATE NOTES

USER

EMAIL

PW

PASSWORD GROUP

NAME DATE NOTES

USER

EMAIL

PW

NAME DATE NOTES

USER

EMAIL

PW

NAME DATE NOTES

USER

EMAIL

PW

NAME DATE NOTES

USER

EMAIL

PW

NAME DATE NOTES

USER

EMAIL

PW

NAME DATE NOTES

USER

EMAIL

PW

NAME DATE NOTES

USER

EMAIL

PW

NAME DATE NOTES

USER

EMAIL

PW

NAME DATE NOTES

USER

EMAIL

PW

NAME DATE NOTES

USER

EMAIL

PW

NAME DATE NOTES

USER

EMAIL

PW

NAME DATE NOTES

USER

EMAIL

PW

NAME DATE NOTES

USER

EMAIL

PW

NAME DATE NOTES

USER

EMAIL

PW

NAME DATE NOTES

USER

EMAIL

PW

NAME DATE NOTES

USER

EMAIL

PW

NAME DATE NOTES

USER

EMAIL

PW

NAME DATE NOTES

USER

EMAIL

PW

NAME DATE NOTES

USER

EMAIL

PW

NAME DATE NOTES

USER

EMAIL

PW

PASSWORD GROUP

NAME DATE NOTES

USER

EMAIL

PW

NAME DATE NOTES

USER

EMAIL

PW

NAME DATE NOTES

USER

EMAIL

PW

NAME DATE NOTES

USER

EMAIL

PW

NAME DATE NOTES

USER

EMAIL

PW

NAME DATE NOTES

USER

EMAIL

PW

NAME DATE NOTES

USER

EMAIL

PW

NAME DATE NOTES

USER

EMAIL

PW

NAME DATE NOTES

USER

EMAIL

PW

NAME DATE NOTES

USER

EMAIL

PW

NAME DATE NOTES

USER

EMAIL

PW

NAME DATE NOTES

USER

EMAIL

PW

NAME | DATE | NOTES

USER

EMAIL

PW

NAME | DATE | NOTES

USER

EMAIL

PW

NAME | DATE | NOTES

USER

EMAIL

PW

NAME | DATE | NOTES

USER

EMAIL

PW

NAME DATE NOTES

USER

EMAIL

PW

NAME DATE NOTES

USER

EMAIL

PW

NAME DATE NOTES

USER

EMAIL

PW

NAME DATE NOTES

USER

EMAIL

PW

PASSWORD GROUP

NAME DATE NOTES

USER

EMAIL

PW

NAME DATE NOTES

USER

EMAIL

PW

NAME DATE NOTES

USER

EMAIL

PW

NAME DATE NOTES

USER

EMAIL

PW

NAME DATE NOTES

USER

EMAIL

PW

NAME DATE NOTES

USER

EMAIL

PW

NAME DATE NOTES

USER

EMAIL

PW

NAME DATE NOTES

USER

EMAIL

PW

NAME DATE NOTES

USER

EMAIL

PW

NAME DATE NOTES

USER

EMAIL

PW

NAME DATE NOTES

USER

EMAIL

PW

NAME DATE NOTES

USER

EMAIL

PW

NAME DATE NOTES

USER

EMAIL

PW

NAME DATE NOTES

USER

EMAIL

PW

NAME DATE NOTES

USER

EMAIL

PW

NAME DATE NOTES

USER

EMAIL

PW

NAME DATE NOTES

USER

EMAIL

PW

NAME DATE NOTES

USER

EMAIL

PW

NAME DATE NOTES

USER

EMAIL

PW

NAME DATE NOTES

USER

EMAIL

PW

PASSWORD GROUP

NAME	DATE	NOTES
USER		
EMAIL		
PW		

NAME	DATE	NOTES
USER		
EMAIL		
PW		

NAME	DATE	NOTES
USER		
EMAIL		
PW		

NAME	DATE	NOTES
USER		
EMAIL		
PW		

NAME DATE NOTES

USER

EMAIL

PW

NAME DATE NOTES

USER

EMAIL

PW

NAME DATE NOTES

USER

EMAIL

PW

NAME DATE NOTES

USER

EMAIL

PW

NAME DATE NOTES

USER

EMAIL

PW

NAME DATE NOTES

USER

EMAIL

PW

NAME DATE NOTES

USER

EMAIL

PW

NAME DATE NOTES

USER

EMAIL

PW

NAME DATE NOTES

USER

EMAIL

PW

NAME DATE NOTES

USER

EMAIL

PW

NAME DATE NOTES

USER

EMAIL

PW

NAME DATE NOTES

USER

EMAIL

PW

NAME DATE NOTES

USER

EMAIL

PW

NAME DATE NOTES

USER

EMAIL

PW

NAME DATE NOTES

USER

EMAIL

PW

NAME DATE NOTES

USER

EMAIL

PW

PASSWORD GROUP

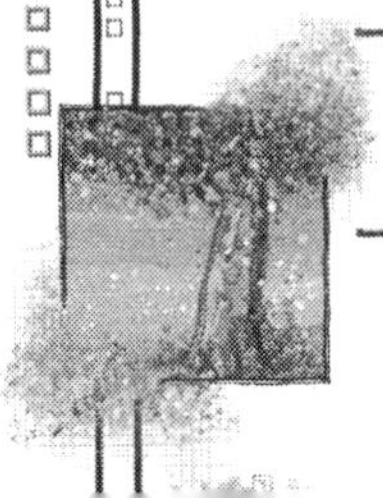

NAME DATE NOTES

USER

EMAIL

PW

NAME DATE NOTES

USER

EMAIL

PW

NAME DATE NOTES

USER

EMAIL

PW

NAME DATE NOTES

USER

EMAIL

PW

NAME DATE NOTES

USER

EMAIL

PW

NAME DATE NOTES

USER

EMAIL

PW

NAME DATE NOTES

USER

EMAIL

PW

NAME DATE NOTES

USER

EMAIL

PW

NAME DATE NOTES

USER

EMAIL

PW

NAME DATE NOTES

USER

EMAIL

PW

NAME DATE NOTES

USER

EMAIL

PW

NAME DATE NOTES

USER

EMAIL

PW

NAME DATE NOTES

USER

EMAIL

PW

NAME DATE NOTES

USER

EMAIL

PW

NAME DATE NOTES

USER

EMAIL

PW

NAME DATE NOTES

USER

EMAIL

PW

NAME

DATE

NOTES

USER

EMAIL

PW

NAME

DATE

NOTES

USER

EMAIL

PW

NAME

DATE

NOTES

USER

EMAIL

PW

NAME

DATE

NOTES

USER

EMAIL

PW

PASSWORD GROUP

NAME DATE NOTES

USER

EMAIL

PW

NAME DATE NOTES

USER

EMAIL

PW

NAME DATE NOTES

USER

EMAIL

PW

NAME DATE NOTES

USER

EMAIL

PW

NAME DATE NOTES

USER

EMAIL

PW

NAME DATE NOTES

USER

EMAIL

PW

NAME DATE NOTES

USER

EMAIL

PW

NAME DATE NOTES

USER

EMAIL

PW

NAME DATE NOTES

USER

EMAIL

PW

NAME DATE NOTES

USER

EMAIL

PW

NAME DATE NOTES

USER

EMAIL

PW

NAME DATE NOTES

USER

EMAIL

PW

NAME DATE NOTES

USER

EMAIL

PW

NAME DATE NOTES

USER

EMAIL

PW

NAME DATE NOTES

USER

EMAIL

PW

NAME DATE NOTES

USER

EMAIL

PW

NAME DATE NOTES

USER

EMAIL

PW

NAME DATE NOTES

USER

EMAIL

PW

NAME DATE NOTES

USER

EMAIL

PW

NAME DATE NOTES

USER

EMAIL

PW

PASSWORD GROUP

NAME DATE NOTES

USER

EMAIL

PW

NAME DATE NOTES

USER

EMAIL

PW

NAME DATE NOTES

USER

EMAIL

PW

NAME DATE NOTES

USER

EMAIL

PW

NAME DATE NOTES

USER

EMAIL

PW

NAME DATE NOTES

USER

EMAIL

PW

NAME DATE NOTES

USER

EMAIL

PW

NAME DATE NOTES

USER

EMAIL

PW

NAME DATE NOTES

USER

EMAIL

PW

NAME DATE NOTES

USER

EMAIL

PW

NAME DATE NOTES

USER

EMAIL

PW

NAME DATE NOTES

USER

EMAIL

PW

NAME DATE NOTES

USER

EMAIL

PW

NAME DATE NOTES

USER

EMAIL

PW

NAME DATE NOTES

USER

EMAIL

PW

NAME DATE NOTES

USER

EMAIL

PW

NAME DATE NOTES

USER

EMAIL

PW

NAME DATE NOTES

USER

EMAIL

PW

NAME DATE NOTES

USER

EMAIL

PW

NAME DATE NOTES

USER

EMAIL

PW

PASSWORD GROUP

NAME DATE NOTES

USER

EMAIL

PW

NAME DATE NOTES

USER

EMAIL

PW

NAME DATE NOTES

USER

EMAIL

PW

NAME DATE NOTES

USER

EMAIL

PW

NAME DATE NOTES

USER

EMAIL

PW

NAME DATE NOTES

USER

EMAIL

PW

NAME DATE NOTES

USER

EMAIL

PW

NAME DATE NOTES

USER

EMAIL

PW

NAME DATE NOTES

USER

EMAIL

PW

NAME DATE NOTES

USER

EMAIL

PW

NAME DATE NOTES

USER

EMAIL

PW

NAME DATE NOTES

USER

EMAIL

PW

NAME DATE NOTES

USER

EMAIL

PW

NAME DATE NOTES

USER

EMAIL

PW

NAME DATE NOTES

USER

EMAIL

PW

NAME DATE NOTES

USER

EMAIL

PW

NAME DATE NOTES

USER

EMAIL

PW

NAME DATE NOTES

USER

EMAIL

PW

NAME DATE NOTES

USER

EMAIL

PW

NAME DATE NOTES

USER

EMAIL

PW

PASSWORD GROUP

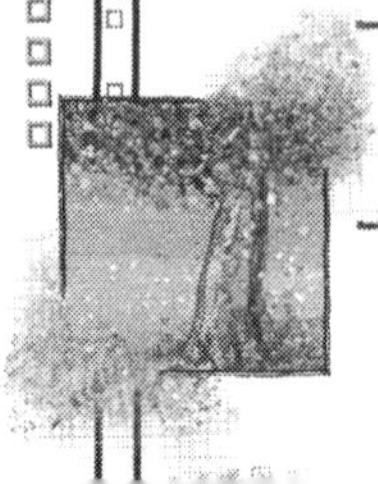

NAME DATE NOTES

USER

EMAIL

PW

NAME DATE NOTES

USER

EMAIL

PW

NAME DATE NOTES

USER

EMAIL

PW

NAME DATE NOTES

USER

EMAIL

PW

NAME DATE NOTES

USER

EMAIL

PW

NAME DATE NOTES

USER

EMAIL

PW

NAME DATE NOTES

USER

EMAIL

PW

NAME DATE NOTES

USER

EMAIL

PW

NAME DATE NOTES

USER

EMAIL

PW

NAME DATE NOTES

USER

EMAIL

PW

NAME DATE NOTES

USER

EMAIL

PW

NAME DATE NOTES

USER

EMAIL

PW

NAME DATE NOTES

USER

EMAIL

PW

NAME DATE NOTES

USER

EMAIL

PW

NAME DATE NOTES

USER

EMAIL

PW

NAME DATE NOTES

USER

EMAIL

PW

NAME DATE NOTES

USER

EMAIL

PW

NAME DATE NOTES

USER

EMAIL

PW

NAME DATE NOTES

USER

EMAIL

PW

NAME DATE NOTES

USER

EMAIL

PW

PASSWORD GROUP

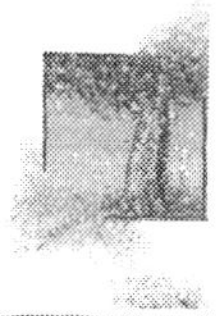

NAME DATE NOTES

USER

EMAIL

PW

NAME DATE NOTES

USER

EMAIL

PW

NAME DATE NOTES

USER

EMAIL

PW

NAME DATE NOTES

USER

EMAIL

PW

NAME DATE NOTES

USER

EMAIL

PW

NAME DATE NOTES

USER

EMAIL

PW

NAME DATE NOTES

USER

EMAIL

PW

NAME DATE NOTES

USER

EMAIL

PW

NAME DATE NOTES

USER

EMAIL

PW

NAME DATE NOTES

USER

EMAIL

PW

NAME DATE NOTES

USER

EMAIL

PW

NAME DATE NOTES

USER

EMAIL

PW

NAME DATE NOTES

USER

EMAIL

PW

NAME DATE NOTES

USER

EMAIL

PW

NAME DATE NOTES

USER

EMAIL

PW

NAME DATE NOTES

USER

EMAIL

PW

NAME DATE NOTES

USER

EMAIL

PW

NAME DATE NOTES

USER

EMAIL

PW

NAME DATE NOTES

USER

EMAIL

PW

NAME DATE NOTES

USER

EMAIL

PW

PASSWORD GROUP

NAME | DATE | NOTES

USER

EMAIL

PW

NAME | DATE | NOTES

USER

EMAIL

PW

NAME | DATE | NOTES

USER

EMAIL

PW

NAME | DATE | NOTES

USER

EMAIL

PW

NAME DATE NOTES

USER

EMAIL

PW

NAME DATE NOTES

USER

EMAIL

PW

NAME DATE NOTES

USER

EMAIL

PW

NAME DATE NOTES

USER

EMAIL

PW

NAME DATE NOTES

USER

EMAIL

PW

NAME DATE NOTES

USER

EMAIL

PW

NAME DATE NOTES

USER

EMAIL

PW

NAME DATE NOTES

USER

EMAIL

PW

NAME DATE NOTES

USER

EMAIL

PW

NAME DATE NOTES

USER

EMAIL

PW

NAME DATE NOTES

USER

EMAIL

PW

NAME DATE NOTES

USER

EMAIL

PW

NAME DATE NOTES

USER

EMAIL

PW

NAME DATE NOTES

USER

EMAIL

PW

NAME DATE NOTES

USER

EMAIL

PW

NAME DATE NOTES

USER

EMAIL

PW

PASSWORD GROUP

NAME DATE NOTES

USER

EMAIL

PW

NAME DATE NOTES

USER

EMAIL

PW

NAME DATE NOTES

USER

EMAIL

PW

NAME DATE NOTES

USER

EMAIL

PW

NAME DATE NOTES

USER

EMAIL

PW

NAME DATE NOTES

USER

EMAIL

PW

NAME DATE NOTES

USER

EMAIL

PW

NAME DATE NOTES

USER

EMAIL

PW

NAME DATE NOTES

USER

EMAIL

PW

NAME DATE NOTES

USER

EMAIL

PW

NAME DATE NOTES

USER

EMAIL

PW

NAME DATE NOTES

USER

EMAIL

PW

NAME DATE NOTES

USER

EMAIL

PW

NAME DATE NOTES

USER

EMAIL

PW

NAME DATE NOTES

USER

EMAIL

PW

NAME DATE NOTES

USER

EMAIL

PW

NAME DATE NOTES

USER

EMAIL

PW

NAME DATE NOTES

USER

EMAIL

PW

NAME DATE NOTES

USER

EMAIL

PW

NAME DATE NOTES

USER

EMAIL

PW

PASSWORD GROUP

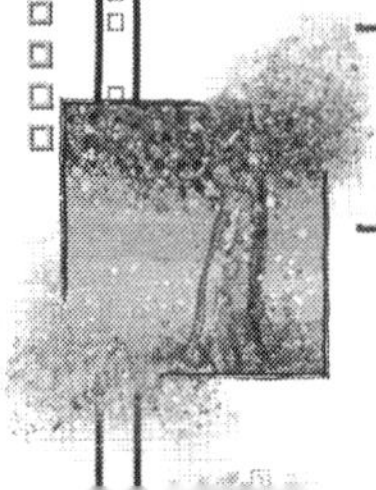

NAME DATE NOTES

USER

EMAIL

PW

NAME DATE NOTES

USER

EMAIL

PW

NAME DATE NOTES

USER

EMAIL

PW

NAME DATE NOTES

USER

EMAIL

PW

NAME DATE NOTES

USER

EMAIL

PW

NAME DATE NOTES

USER

EMAIL

PW

NAME DATE NOTES

USER

EMAIL

PW

NAME DATE NOTES

USER

EMAIL

PW

NAME DATE NOTES

USER

EMAIL

PW

NAME DATE NOTES

USER

EMAIL

PW

NAME DATE NOTES

USER

EMAIL

PW

NAME DATE NOTES

USER

EMAIL

PW

NAME DATE NOTES

USER

EMAIL

PW

NAME DATE NOTES

USER

EMAIL

PW

NAME DATE NOTES

USER

EMAIL

PW

NAME DATE NOTES

USER

EMAIL

PW

NAME DATE NOTES

USER

EMAIL

PW

NAME DATE NOTES

USER

EMAIL

PW

NAME DATE NOTES

USER

EMAIL

PW

NAME DATE NOTES

USER

EMAIL

PW

PASSWORD GROUP

NAME DATE NOTES

USER

EMAIL

PW

NAME DATE NOTES

USER

EMAIL

PW

NAME DATE NOTES

USER

EMAIL

PW

NAME DATE NOTES

USER

EMAIL

PW

NAME DATE NOTES

USER

EMAIL

PW

NAME DATE NOTES

USER

EMAIL

PW

NAME DATE NOTES

USER

EMAIL

PW

NAME DATE NOTES

USER

EMAIL

PW

NAME DATE NOTES

USER

EMAIL

PW

NAME DATE NOTES

USER

EMAIL

PW

NAME DATE NOTES

USER

EMAIL

PW

NAME DATE NOTES

USER

EMAIL

PW

NAME DATE NOTES

USER

EMAIL

PW

NAME DATE NOTES

USER

EMAIL

PW

NAME DATE NOTES

USER

EMAIL

PW

NAME DATE NOTES

USER

EMAIL

PW

NAME		DATE	NOTES
USER			☐ ☐
EMAIL			☐ ☐
PW			☐ ☐

NAME		DATE	NOTES
USER			☐ ☐
EMAIL			☐ ☐
PW			☐ ☐

NAME		DATE	NOTES
USER			☐ ☐
EMAIL			☐ ☐
PW			☐ ☐

NAME		DATE	NOTES
USER			☐ ☐
EMAIL			☐ ☐
PW			☐ ☐

PASSWORD GROUP

NAME DATE NOTES

USER

EMAIL

PW

NAME DATE NOTES

USER

EMAIL

PW

NAME DATE NOTES

USER

EMAIL

PW

NAME DATE NOTES

USER

EMAIL

PW

NAME DATE NOTES

USER

EMAIL

PW

NAME DATE NOTES

USER

EMAIL

PW

NAME DATE NOTES

USER

EMAIL

PW

NAME DATE NOTES

USER

EMAIL

PW

NAME DATE NOTES

USER

EMAIL

PW

NAME DATE NOTES

USER

EMAIL

PW

NAME DATE NOTES

USER

EMAIL

PW

NAME DATE NOTES

USER

EMAIL

PW

NOTES & IDEAS

If there is any feature you would suggest for the next version of this journal, what would it be?

These pages are for any notes you need to make.

Remember to schedule password changes regularly.

WORKSHEET LIST

Use this checklist for any tasks you need to complete as you set up and maintain this password keeper logbook.

WORKSHEET LIST

Use this checklist for any tasks you need to complete as you set up and maintain this password keeper logbook.

- [ ]
- [ ]
- [ ]
- [ ]
- [ ]
- [ ]
- [ ]
- [ ]
- [ ]
- [ ]
- [ ]
- [ ]
- [ ]
- [ ]
- [ ]
- [ ]

- [ ]
- [ ]
- [ ]
- [ ]
- [ ]
- [ ]
- [ ]
- [ ]
- [ ]
- [ ]
- [ ]
- [ ]
- [ ]
- [ ]
- [ ]
- [ ]

WORKSHEET LIST

Use this checklist for any tasks you need to complete as you set up and maintain this password keeper logbook.

WORKSHEET LIST

Use this checklist for any tasks you need to complete as you set up and maintain this password keeper logbook.

- [ ]
- [ ]
- [ ]
- [ ]
- [ ]
- [ ]
- [ ]
- [ ]
- [ ]
- [ ]
- [ ]
- [ ]
- [ ]
- [ ]
- [ ]
- [ ]
- [ ]
- [ ]
- [ ]
- [ ]
- [ ]
- [ ]
- [ ]
- [ ]
- [ ]
- [ ]
- [ ]
- [ ]
- [ ]
- [ ]
- [ ]
- [ ]
- [ ]
- [ ]

journaling scribbles

A Series of Color-Coded Journals for Creativity & Planning

Look for additional designs!

Each Journal in the Series
Fits into One or More of the 9 Categories.

Which journal will you choose next?

1. planners
2. Writing
3. Spiritual Life
4. Family & Friends
5. HOUSE & THINGS
6. adventures
7. Creativity
8. Projects & Activities
9. wellness

oh, the people i've met
through the formation of these journals!

A special thanks to my endearing husband who offers his puns, humor, encouragement & keen insights. He's also a poet!

To Terha Knittel, lovely composer and vocalist extraordinaire, who composed the music for video trailers and promotions.

To Kyle Knittel who guided me through video production.

To Erica Alexander who formatted the 3D covers of my artwork.

To Wendy Hibbard & Alisa Hope Wagner who gave advice on formatting interiors and offered publishing war stories.

To my launch team on FB who helped edit, proofread, and cheer me toward publication. My gratitude to these idea generators, amazing brainstorming networkers, and the many cheerleaders, thought-provoking mentors, trouble shooters, formatting ninjas, and "let-it-be-enough" believers, trust soldiers, and prayer warriors.

Thank you for blessing my proverbial socks off!

To Deena Adams, Erica Alexander, Johnnie Alexander, Starr Ayers, Cathy Baker, Shelly Beach, Bonnie Beardsley, Kathy Bruins, Diane Buie, Rachael M. Colby, Teresa Dunn, Annette Marie Griffin, Ginger Harrington, Melissa Henderson, Denise Heathman, Debra Kerr Hofland, Alisha Holt, Monica Huyser, Heather Kaufman, Mary Potter Kenyon, Terha Knittel, Sue Kuenzi, Keren Herrera Lyles, Shelly Mendenhall, Emily Mendoza, Alycia Morales, Teresa Moyer, Diana Parson, Emily Potter, Diana Rockwell, Jodi Saline, Wanda Sanchez, Lyneta Smith, Joleen Steel, Patti Stockdale, Debra DuPree Williams, Shannan L. Williams Schreiner, Adria Wilkins, Imogene Steinbronn Wilkinson, and others who may yet join me. If have forgotten anyone, please forgive me and let me know so I can add you to the next journal.

Blessings to each of you!

Enjoy a bonus 30 day writing prompt on the next page.

what will you write next?

Guided Journal Writing Prompts for an Additional 30 days:

1-2 I'm a child of God wrapped in a blanket of courage and faith.

3-4 I'm an eclectic creative, spinning patterns of joy.

5-6 I'm an artist woven in textures of beauty & hope.

7-8 I'm a song of direction, ordering steps, like the notes on a page, or words left to right.

9-10 I'm a puzzle, putting pieces together & solving mysteries.

11-12 I'm a networking ninja, making intentional connections.

13-14 I'm a gardener planting seeds and growing flowers.

15-16 I'm an architect and builder, building strong foundations.

17-18 I'm a prayer warrior, falling on my knees and lifting prayers to God.

19-20 I'm brainstormer, mapping ideas to problems and finding solutions.

21-22 I'm a writer, discipling and cultivating grace.

23-24 I'm no one, and yet, I'm enough,

25-26 I'm set apart for His glory alone.

27-28 Thank you, Lord, for your salvation.

29-30 Thank you, Lord, for giving me breath.

please know that i am praying for you.

I'd love to hear from you. If you have a journal page to share with me, Tag me on IG @PatriciaTiffanyMorris or on FB @TiffanyInksStudio

Made in United States
Orlando, FL
04 November 2026

90205463R00088